U.S. Department of Justice
Office of Justice Programs
National Institute of Justice

I0475750

Research Report

Case Management Reduces Drug Use and Criminality Among Drug-Involved Arrestees: An Experimental Study of an HIV Prevention Intervention

National Institute of Justice

National Institute on Drug Abuse

Foreword

The spread of HIV/AIDS has lent new urgency to the issue of substance abuse because of the high risk for transmission posed through injection drug use and sexual contact with individuals who themselves have injected drugs. With the likelihood of arrest far greater among heavy drug users than among people who do not use drugs, a critical public health concern is also a concern of criminal justice.

Reducing substance abuse and the high-risk behaviors associated with injecting drugs is a key part of any strategy to prevent HIV/AIDS. The study findings presented here suggest a promising approach for assisting arrestees who use illicit drugs and are at risk for becoming infected with HIV. The approach the researchers found to be successful is case management—intensive intervention involving a full spectrum of services from assessment and treatment planning through counseling, monitoring, and advocacy. Case management not only reduced drug use and increased the use of substance abuse treatment among the drug-involved arrestees tested, but at the same time lowered recidivism. That the approach can be used to effect in a criminally involved population without resort to coercion is an equally important finding. And although it was not as successful in reducing high-risk behavior related to sexual practices and injection drug use, the researchers suggest that further refinements to the model defined in this study may lead to improved outcomes in high-risk behaviors.

This report reflects an important linkage of both NIJ's and NIDA's interests in integrating critical public safety and public health approaches in working with drug-involved arrestees. The drug-crime nexus has joined NIDA and NIJ in a number of projects in the past, and our longstanding association has served as the foundation for current efforts to develop mutual research priorities in the area of drug-involved offenders. The study presented here amply demonstrates the wisdom and utility of such collaborative efforts, and we will continue to pursue them. We also would encourage further study of the approach explored here. The association of injection drug use with a considerable proportion of AIDS cases and HIV infections, and the promise held by the case management approach, make it imperative that such research be conducted and supported.

Jeremy Travis
Director
National Institute of Justice

Alan I. Leshner
Director
National Institute on Drug Abuse

Case Management Reduces Drug Use and Criminality Among Drug-Involved Arrestees: An Experimental Study of an HIV Prevention Intervention

William Rhodes

Michael Gross

A Final Summary Report Presented to the National Institute of Justice and the National Institute on Drug Abuse

March 1997

U.S. Department of Justice
Office of Justice Programs

National Institute of Justice
Jeremy Travis
Director

Cheryl Crawford
Project Monitor

U.S. Department of Health and Human Services
National Institute on Drug Abuse
Alan I. Leshner
Director

This project was supported under cooperative agreement with Abt Associates Inc. (89–IJ–CX–0060) with support from the National Institute of Justice, Office of Justice Programs, U.S. Department of Justice, and the National Institute on Drug Abuse, U.S. Department of Health and Human Services. The authors of this report, William Rhodes and Michael Gross, are senior scientists at Abt Associates Inc. at its Cambridge, Massachusetts, and Bethesda, Maryland, offices, respectively. Opinions or points of view expressed in this document are those of the authors and do not necessarily reflect the official position of the U.S. Department of Justice or the U.S. Department of Health and Human Services.

NCJ 155281

The National Institute of Justice is a component of the Office of Justice Programs, which also includes the Bureau of Justice Assistance, Bureau of Justice Statistics, Office of Juvenile Justice and Delinquency Prevention, and the Office for Victims of Crime.

Table of Contents

Acknowledgments

The authors gratefully acknowledge the sustained and valuable guidance provided by members of the Federal Advisory Board for the cooperative agreement under which this research was conducted. Cheryl Crawford, of the National Institute of Justice (NIJ), was unstintingly helpful and consistently insightful. Other members of the Federal Advisory Board, whose contributions account for many of the strengths of this project and for none of its errors of omission or commission, include: Virginia Baldau, NIJ; Robert Battjes, the National Institute on Drug Abuse (NIDA); Barry S. Brown, University of North Carolina; Paul Cascarano, NIJ; Peter Delany, NIDA; Carl Leukefeld, University of Kentucky; Arnold Mills, NIDA; and Eric Wish, Center for Substance Abuse Research, University of Maryland.

For their substantial involvement in the implementation of this project and the analysis of its findings, the authors are especially indebted to a number of colleagues at Abt Associates Inc. who are coauthors of the final project report: Cathy Conly, Tammy Enos, Theresa Mason, Stacia Langenbahn, and Linda Truitt.

Introduction

Recent findings from a research study conducted with support from the National Institute of Justice (NIJ) and the National Institute on Drug Abuse (NIDA) suggest that intensive case management, delivered for 6 months, can reduce drug use and recidivism and increase use of substance abuse treatment among drug-involved arrestees released after booking. This study, which was based on a controlled experiment involving close to 1,400 arrestees, was conducted between August 1991 and April 1993 in Washington, D.C., and Portland, Oregon.

Case management also was intended to reduce sexual- and injection-related risk behaviors implicated in the transmission of human immunodeficiency virus (HIV), the causative agent of acquired immunodeficiency syndrome (AIDS). Case management was less successful in improving these specific HIV-related outcomes. However, the study suggests how further refinements to the case management model might elicit improved outcomes associated with HIV prevention.

Particularly relevant to the criminal justice context, case management as provided in this study was strictly voluntary. High levels of participation were sustained among participants in the study without either criminal sanctions for noncompliance or material rewards for participation.

Statement of the Problem

By the time the intervention phase of this study ended in 1992, injection drug use had accounted for nearly a quarter of all adult cases of AIDS and half of all cases diagnosed among women (CDC, 1993a). Heterosexual transmission had begun to account for the largest proportional increase in reported AIDS cases (CDC, 1994). Most heterosexually acquired HIV infections had resulted from sexual contact with persons who themselves became infected through injection drug use. Moreover, almost all infants in the United States born with HIV infection had been exposed because their mother acquired HIV through drug injection or sexual contact with an injection drug user.

HIV prevention efforts restricted to injection drug users (IDUs), rather than offered to the broader population of drug users, miss an opportunity to intervene before users progress to or resume injection (Des Jarlais and Friedman, 1987). Focusing exclusively on injectors also does not address the significant potential for sexual transmission among drug-involved noninjectors (Stimson, 1992; Leigh and Stall, 1993; Steel and Haverkos, 1992; Weissman et al., 1991). For example, compulsive female crack users often consent to unprotected sex with multiple partners in exchange for money or drugs (Steel and Haverkos, 1992; Weissman et al., 1991).

For these reasons, this study targeted users of illicit drugs other than marijuana, whether or not they had a history of injection drug use: a population readily found in criminal justice settings. People who use drugs heavily are more likely to be arrested than people who do not use drugs.[1] Other than entitlement programs, jail and prison are the public institutions most likely to process both male and female IDUs who are not in treatment (Gross and Brown, 1993). Thus, the criminal justice system (CJS) seems to be an ideal place to situate an HIV prevention intervention (Wish, O'Neil, and Baldau, 1990; Leukefeld, Battjes, and Pickens, 1991). Despite the opportunity afforded by the large at-risk population filtering through the CJS, correctional agencies have limited ability to take proactive steps to reduce the spread of HIV (Hammett et al., 1991). Most arrestees are released within hours of arrest. Those convicted often avoid jail and prison where they might be exposed to HIV prevention programs.

Although the period following arrest is an attractive focus for an intervention because it is the time when the recruitment pool is largest, interventions with an arrested population pose procedural problems. An intensive intervention cannot be carried out in the jail environment because facilities are seldom designed for either group education sessions or confidential interviews. Booking and release usually occur within 2 or 3 days, and often on the same day. During that period, arrestees often are inebriated or in withdrawal and otherwise suffering from the stress of incarceration. Furthermore, some topics seem unsuitable for a jail environment. Teaching needle-cleaning techniques or safe sex practices runs counter to the criminal justice message that drug use and prostitution are illegal. As a result, lockups and booking facilities were used in this study as points of contact with prospective participants, but the interventions occurred outside the CJS. The overall sequence of events is depicted in figure 1 and described more fully below.

Design of the study
While heavy drug users and their sexual partners seem to know how to prevent exposure to HIV, other factors appear to militate against behavioral change (Gross et al., 1992; Hammett et al., 1992). The interventions employed in this study focused on two factors believed to be associated with behavioral change: (a) social support, including perceptions of peer norms favoring risk reduction and encouragement from others to change behavior; and (b) removal of practical barriers that might discourage behavioral change. For reasons detailed below, case management appeared to be a plausible strategy for addressing these considerations. The study addressed the following research question: Would case management, as implemented in this study, reduce high-risk behaviors by drug users recruited during the period of pretrial release?

Rationale for case management as an enhanced intervention
At the time this study was designed, researchers had reported that persons at risk for exposure to HIV often cited their preoccupation with more exigent and multiple life needs (for example, job, housing/shelter, and drug treatment) to explain why they were not dealing with HIV prevention (Finn et al., 1992). Based on similar observations about impediments to behavioral change, other researchers had recommended case

management as a strategy for addressing drug use (Martin and Scarpitti, 1993; McLellan et al., 1993; Siegal, 1994) and HIV risk behavior (CDC, 1993b; Ashery et al., 1992).

Substance abusers face unique barriers to receiving services. They have the reputation of being "the least desirable group with which to work, the most unstable, the most uncooperative, and the least understood; and some programs have specifically eliminated substance abusers by their eligibility criteria" (Ashery et al., 1992). Philosophical conflicts over whether drug addiction is a choice (deserving of punitive consequences) or a disease (requiring treatment), and the association of addiction with criminality also can impede clients' access to services (Ridgely and Willenbring, 1992). Because of its focus on leveraging difficult-to-access services, case management is particularly appropriate for persons with both criminal and drug involvement.

A final and compelling reason for using case management in this experiment was a practical response to constraints inherent in the criminal justice context. In the political and historical context in which this study was designed, many criminal justice agencies might have resisted sponsoring interventions that implicitly tolerated illicit drug use by providing instruction in needle hygiene. Even messages advocating condom use were problematic in their implication that abstinence or mutual monogamy was not normative. Case management, emphasizing referrals to community agencies that could deliver comprehensive HIV prevention counseling unfettered by such constraints, seemed likely to be acceptable across a variety of criminal justice settings. Case management also seemed to be the most cost-effective means of making services supportive of HIV risk reduction available to the target population.

Case management typically combines the following components: assessment, treatment planning, linkage, referrals, monitoring, and advocacy (Weil et al., 1985; Anthony et al., 1988; Applebaum and Austin, 1990). It may also include elements of counseling, therapy, or social support (Rothman, 1991). Traditional probation and parole includes many components associated with case management (Rothman, 1971) such as assessment, planning, linkage, and advocacy. However, controlling illegal behaviors— the principal function of probation, parole, and other forms of community supervision—played no role in the form of case management used in this study.

Approach to case management in this study

When the study was designed, much had been written about case management. However, little was known about how to configure case management specifically to serve a drug-involved and criminally active population without coercing participation, or about how to offer case management as a means of fostering HIV risk reduction. For the design of key elements of the programs, reliance was placed on the experience and judgment of the agencies contracted to implement the study in Portland and in Washington, D.C. These agencies had provided case management to similar populations for decades, albeit in a criminal justice framework that included sanctions for nonparticipation.

The contracting agencies were expected to hire appropriate case managers; to identify qualified supervisory staff and arrange for staff training; to broker and monitor access to relevant services; and to provide adequate protocols and forms for conducting needs assessment, treatment planning, and clinical monitoring. The sites were allowed considerable latitude in implementing case management consistent with their understanding of best practices. (Appendix A describes in more detail key features of case management as implemented in this study.) Nevertheless, certain minimal parameters were stipulated, including:

■   Average caseload size was set at 30 per full-time case manager. An average of two face-to-face contacts and two telephone contacts per month was established as a minimum level of service for each active client.

■   Categories of community service providers were specified with which sites were required to negotiate formal referral arrangements and agreements. These included drug treatment programs, HIV counseling and testing sites, HIV prevention programs, and other key health and human service agencies (e.g., job training and employment counseling and medical clinics).

■   Case management and other study staff were prohibited from providing information about any project participant to the criminal justice system.

In hiring case managers, sites were encouraged to emphasize an ability to empathize with the client population targeted for the study while encouraging favorable behavior changes. They were advised to seek personnel with comparable racial/ethnic and cultural backgrounds. Educational and experience credentials were not specified, based on the assumption that specific skill deficits would be addressed by inservice training and continuing supervision. Half of the case managers in Portland, Oregon, and all case managers in Washington, D.C., had bachelor's degrees; the remainder had completed high school. Many had traditional social service experience in social work, probation, and drug counseling. Typically, however, their experience was limited in scope or duration. Others came with hands-on experience in crisis intervention, AIDS education, or client advocacy. Three of nine case managers in Portland and one of five in Washington, D.C., acknowledged histories of drug or alcohol abuse; all had been in recovery less than 5 years. In essence, the case managers were more like paraprofessionals than professional case workers.[2]

Because "compliance" was to be strictly voluntary, responsibility for clients' participation in case management and for their use of referrals rested with the case managers. That is, case managers were told that a critical part of their job was to engage clients in the case management process and to motivate them to seek help from appropriate sources without coercion or significant financial incentives. All participants, no matter what their experimental assignment, were offered a total of $60 for completing three interviews. However, participants assigned to case management knew they would be paid for the interviews whether or not they met with their case managers. Bus tokens were offered on an as-needed basis for visits to the project site. Case managers also had limited funds that could be used on rare occasions to buy a client lunch; snacks and beverages were available at the project offices.

Case managers were encouraged to be creative and were afforded wide latitude about the methods they could use, including meeting clients at their home or elsewhere in the community, going to visit them in jail if necessary, taking them to lunch on occasion, and personally accompanying them to agencies to which they were referred. Program coordinators were encouraged to be flexible about making reassignments without blame if the first case manager assigned did not establish rapport with the client. Case managers were allowed to deliver referrals at varying levels of intensity. This ranged from giving the client a name and phone number, to making an appointment on behalf of the client, to actually accompanying the client to an agency or program to facilitate admission.

Comparison interventions

In order to assess its effectiveness, case management was delivered in the context of an experiment. Participants were assigned at random to one of three interventions:

1. **VIDEO:** In the "control" condition, participants viewed a **videotape** developed specifically for this population and received a **referral guide** to relevant services in their community.

2. **REFERRAL:** In this intermediate intervention, in addition to viewing the **videotape** and receiving the **referral guide**, participants received one **counseling and referral session** with a referral specialist.

3. **CASE MANAGEMENT:** In the enhanced intervention condition, in addition to viewing the **videotape** and receiving the **referral guide**, participants were assigned to the **6-month case management program** described above.

Outcomes were evaluated using formal assessment instruments to measure self-reported behavior at baseline and again at 3 and 6 months. Independent data from criminal justice and drug treatment systems were analyzed to gauge the validity of these self-reports.

Control Intervention: VIDEO	Intermediate Intervention: REFERRAL	Enhanced Intervention: CASE MANAGEMENT

videotape
referral guide
	videotape
	referral guide
	single session with referral
	    specialist
		videotape
		referral guide
		case management
		    (6 months)

An important goal of this study was to produce a videotape to be shown in criminal justice settings, especially lockups and booking facilities, where video playback equipment often is readily available. (The videotape is described more fully in Appendix B.) As noted, all participants viewed the videotape. Many viewed it while in the lockup, and all saw it at the project site. The videotape sought to motivate help-seeking behavior about drug use and HIV prevention by employing health promotion techniques consistent with the health belief model, social marketing principles, and social learning theory. It emphasized the existence of community support for behavior change in the form of drug treatment, support groups, and self-help programs.

This experiment did not intend to evaluate the impact of the videotape (combined with provision of a referral guide to local community programs and services) against no intervention at all. Rather, the videotape and referral guide were considered a minimal intervention provided to all participants with which the referral and case management interventions could be compared.

The referral intervention consisted of a single face-to-face session in which a referral specialist[3] completed a needs assessment and recommended an action plan with a staged sequence of referrals tailored to the participant's needs. The experiment was designed with the belief that the greatest impact would result from a sustained intervention such as 6 months of case management. However, this costly approach involves a significant investment in organizational infrastructure. It seemed plausible that the crisis precipitated by arrest, combined with the motivational message of the videotape, might predispose drug-involved arrestees to make effective use of a one-time encounter with a referral specialist who could direct them to appropriate community service providers. The referral intervention was included, therefore, to learn whether such a "jump start" would have a sustained and significant impact at much less cost and with much less organizational adaptation than full-fledged case management.[4]

Selection of sites for study implementation

Potential sites were screened to select jurisdictions where the courts, sheriff, and other criminal justice authorities would cooperate with the study. Cooperation meant refraining from interfering with recruitment or random assignment. The refusal of the courts to permit studies involving random assignment (e.g., Siegal and Cole, 1993), or other forms of resistance to studies involving random assignment (e.g., Wexler et al., 1994; Martin and Scarpitti, 1992), have undermined attempts to conduct experimental studies in similar settings.

Cooperation also meant refraining from seeking sites' assistance with criminal justice system supervisory functions. Planning discussions with prospective sites addressed concerns about the project's strict separation from CJS operations. Although the intervention programs were to recruit from the CJS, they were not to provide the CJS with any information about their clients. Local officials were concerned about this prohibition against releasing information about clients' positive accomplishments to CJS authorities. Their concern was addressed by pointing out that referral agencies could provide such information. More difficult for the sites to accept, although they did, were prohibitions against using sanctions to coerce or material incentives to induce program participation. Eventually, both project sites received assurances from the judiciary, from prosecuting authorities, and from public defenders that project records could not be subject to subpoena.

An extended site selection process was conducted to document the sites' capacities to deliver on other requirements as well:

■ Sample sizes sufficient to detect program effectiveness.

■ Experience in conducting followup studies or other evidence of an ability to maintain contact with the study population in order to achieve high retention rates.

■ Expertise in providing case management to similar populations and experience in conducting behavioral research.

Service delivery and data collection components were conducted by the Treatment Alternatives to Street Crime (TASC) agency in Portland, Oregon, and the Bureau of Rehabilitation in Washington, D.C. These agencies met the above criteria and, of equal importance, had long-established working relationships with the corresponding pretrial supervision agencies in their respective cities. Interventions were provided through new operational units that had no visible or apparent association with the CJS. The operational units were given new agency titles. Their printed materials, signs, staff titles, and public affiliations did not reference the parent organizations, known in the community as agencies of criminal justice supervision.

In Portland, the TASC agency established the Rose Center, an office located near, but clearly separate from, local court and jail settings. Although TASC staff were available to address difficulties in program operation, the Rose Center functioned independently and was guided by its own policies and regulations. In Washington, D.C., the Bureau of Rehabilitation, a local agency with decades of experience providing services to substance-abusing clients referred from the CJS, established the Community Health Awareness Project (CHAP). CHAP's location was readily accessible to arrestees released from the District's combined court and lockup facility. The offices at both sites were conveniently positioned on public transportation routes. They were designed to make participants feel comfortable, with lounge seating in waiting areas equipped with coffee, sodas, and snacks.

A part-time (approximately 25 percent of a full-time equivalent) researcher experienced in applied social research in criminal justice settings was assigned as an observer at each site.[5] These individuals made impromptu visits, reviewed forms for data quality, inspected case management files for evidence of breadth and depth of service delivery, observed staff meetings, and interacted informally with study staff. Onsite monitoring provided early warning of operational problems such as deficient rates of recruitment, inconsistencies in or misunderstanding of certain items on the interview forms, noteworthy inconsistencies in performance among service delivery staff, and insufficient followup rates.

Experimental procedures
Staff began recruiting substance-abusing arrestees, regardless of their charged offenses, shortly after arrest and arraignment.[6] Arrestees were asked to report to the project site for a full description of the program. They were also advised that they would receive a total of $60 if they completed all three research interviews, in a series of three payments, one per interview. Various methods were used to increase recruitment, including providing maps to the site, providing bus tokens to clients who had arrived at the site, and at times escorting them directly from the lockup to the study site offices. Whether or not they viewed the videotape in the lockup, participants watched it at the project site. After securing participants' informed consent and completing locator information forms, study staff conducted structured interviews with arrestees at baseline and 3 and 6 months later. Staff who delivered services (case management or referral sessions) did not conduct followup interviews with the corresponding project participants.

The interviews elicited information that would show whether behavior improved between the baseline and the 3-month interview, whether improvement was sustained between the 3-month and 6-month interview, and whether improvement was greater for the case management group than for the other two groups. At the conclusion of the baseline interview, participants were randomly assigned to one of the three experimental groups by a staff member who had no information about the substance of the interview or knowledge of the client's level of HIV risk or other factors salient to the intervention (see figure 1).

Process data were captured on case management records (CMRs) completed at every contact (telephone and face-to-face) between study staff and all participants (case management or others, when such contacts occurred). CMRs recorded referral targets, intensity of referral,[7] location of contact, and whether the contact was scheduled in advance. Finally, as previously mentioned, onsite researchers monitored study operations. Other staff members responsible for conducting the evaluation frequently visited the Portland and Washington, D.C., sites.

Data about program participants and similar arrestees who did not participate in the study also were analyzed for validation of the outcome data. For Washington, D.C., these data were criminal histories and drug test data following arrest and release from jail for 7,585 arrestees. The Portland analysis used similar data on criminal histories (n = 1,408) and public substance abuse treatment records (n = 331).

At the conclusion of the study, 33 open-ended case history interviews were completed with case management clients, and, for comparison, 10 indepth interviews were conducted with arrestees assigned to the other two groups. These interviews asked about drug and criminal history; the context and impact of the incident arrest; experiences in the months since study participation ended; and, for case management clients, incentives and disincentives to participation in intensive case management and experiences with case management. Interviewers examined case management clients' case files to compare their descriptions of the process with the case managers' documentation of their work with clients.

Open-ended interviews with all case managers and study staff also were conducted at the end of the study to gain additional insight into philosophies of case management, case management strategies, barriers to service delivery, supports for behavioral change, and attitudes about HIV.

Evidence for the Effectiveness of Case Management

The following discussion of the evaluation findings shows that the target population for the study was successfully enrolled and was essentially representative of the arrestee population in each jurisdiction. The validity and significance of the findings reported here are supported by a number of factors:

■	Comparisons of the study population with the corresponding arrestee population indicate they are essentially comparable or, in some cases, even more severely drug-involved and criminally involved.

- Only 15 percent of the participants could not be interviewed during the followup period. The composition of the groups available for followup did not differ between the case management group and either of the two other assignment groups (video and referral); those who answered the baseline interview did not differ from those who answered the followup interviews.[8]

- A controlled experiment with random assignment to comparison interventions eliminates biases that may result from self-selection into treatment by motivated participants or biased assignment to treatment of promising or especially needy clients by service providers. Tests for random assignment showed that the three intervention groups were essentially indistinguishable on demographic and relevant behavioral variables. Process data, onsite observation, and long-term case study followup interviews consistently indicated that the integrity of random assignment was maintained throughout the study.

- Important outcomes based on self-reports by participants are to some extent corroborated by analyses of criminal justice and substance abuse treatment records from the local jurisdictions.

Salient characteristics of the target population

As shown below, the target population had substantial needs for HIV prevention and other health and human services. Consistent with differences in the drug-involved arrestee populations in the two cities, the profile or project participants differed in certain respects. In Portland, half the participants injected drugs, and many of those injected cocaine; in Washington, D.C., one-fifth of the participants injected drugs, and in almost all instances that drug was heroin. The Portland cohort was somewhat younger than the Washington, D.C., cohort, and a much higher proportion of Portland participants (49 percent versus 29 percent) reported unstable housing situations.

Demographics. At each site (see table 1), approximately 700 arrestees volunteered to participate in the study; three-fourths were men; almost all were between 20 and 40 years old. In Washington, D.C., almost all were African American, as were one-third in Portland. More than one-third of the recruits in both cities had not completed high school.

Study participants for the most part were representative of the overall pool of arrestees in the respective cities. They were slightly more likely to be female: sites were required to recruit samples of women at least proportional to their representation in the arrestee population. They were slightly more likely to be black and older than the general arrestee population. Portland's substantial fraction of Latino arrestees were, at about 7 percent of the sample, underrepresented in comparison with their

12-percent prevalence in the CJS population—even though the study staff included bilingual recruiters and a bicultural case manager.[9] Participants in Washington, D.C., were less educated than typical arrestees. None of these differences were great, and other than gender, they may have resulted from selecting current drug users.

Drug and needle use. By design, enrollment was not restricted to IDUs. Nevertheless, one-fifth of those recruited in Washington and over half of those recruited in Portland were injecting drugs in the month before they were arrested (see table 2). Among current injectors, 23 percent in Washington, D.C., and 39 percent in Portland were sharing injection equipment. More than one-third of the Washington sample had injected at some point in the past, as had two-thirds of the Portland sample. Almost half of those recruited in Washington, D.C., and more than one-third of those recruited in Portland were using either heroin, cocaine, or both at least four times a week at the time of arrest (defined as "heavy use" hereafter in this report).

Table 2 provides a convenient typology of drug use patterns, but, in fact, heroin users sometimes use cocaine and alcohol, and cocaine users sometimes use heroin and alcohol. For example, in Washington, about half the daily heroin injectors routinely added small amounts of cocaine to their heroin use—a common consumption pattern in the era of relatively inexpensive cocaine during which these data were collected. Because their expenditures on cocaine were so small compared with their purchases of heroin, table 2 treats them as heavy heroin users. Similarly, cocaine users occasionally used heroin, but their cocaine purchases were much greater, so table 2 identifies them as cocaine users. Finally, heavy alcohol use is common to both heroin and cocaine users, and table 2 treats heavy alcohol users as a residual category of people who were not otherwise classified as cocaine or heroin users.

Sexual risk. Only a minority (7 percent) of the Washington, D.C., sample had identifiably high-risk partners; another 10 percent were unsure of their partners' level of risk. Consistent with higher rates of drug injection in Portland, 20 percent of the sample had an injection-drug-using partner. Another 6 percent had multiple partners whose level of risk was unknown. Nine percent of the Washington sample and 3 percent of the Portland sample indicated that they exchanged sex for money or drugs.

Half of the sexually active sample in Washington and 60 percent of the sexually active sample in Portland said that during the month before they enrolled in the study they never used condoms with any of their partners. These rates of condom avoidance applied as well to those fractions of the sexually active samples in both cities who either injected drugs or whose sexual partners injected drugs. Consistent with findings in other projects (Turner et al., 1989; Deren et al., 1993), sex workers were most likely to report that they used condoms always (75 percent in Washington, D.C.; 55 percent in Portland), or sometimes (17 percent in Washington, D.C.; 24 percent in Portland).

Criminality. Self-reported criminal activity was similar in both cities. Participants said they had committed on average from 13 to 14 crimes in the month before enrollment, yielding an average income for the month of $450 to $500. Over half of these crimes involved drug dealing.

Service needs. Consistent with expectations, the populations in both cities had multiple needs, especially for employment, health care, counseling, and housing (see table 3).

Delivery of case management

Despite well-justified concern that the target population would decline to participate in a voluntary program such as this (Inciardi et al., 1994), the subgroup assigned to case management participated at a relatively high level. Of the 229 clients assigned to case management in Washington, D.C., only one had no contact with a case manager. Twenty-six percent met or exceeded the original goal of 24 contacts.[10] The majority (62 percent) had two or more contacts of some sort each month (i.e., 13 or more total contacts; see table 4). The five case managers carried an average caseload of 33 clients per case manager, with a range from 30 to 42. In Portland, 94 percent (n = 217) of all case management clients had at least one case management session. Thirty-five percent of the clients had two or more contacts a month (i.e., 13 or more contacts; see table 4). Maximum caseloads ranged from 10 to 40 people per case manager (based on the busiest month for each case manager), with an average caseload of 22.

Both programs were more office based than expected. In Washington, D.C., only 17 percent of all face-to-face contacts with case managers occurred outside the office; the corresponding figure for Portland was 11 percent. A case manager physically accompanied a client to a referral site in only 2 percent of contacts in either program.

In Washington, D.C., almost all clients received at least one referral to drug/alcohol treatment as well as to a self-help program. Services for which referrals were provided with the next greatest frequency (almost three-fourths of all clients) were related to employment and to HIV counseling and testing. Also, in Portland, the highest proportion of referrals (54 percent of all clients) were to drug/alcohol treatment,[11] followed by employment-related (45 percent) and housing-related services (39 percent).

Almost half of all documented contacts between clients and case managers in Washington, D.C., resulted in no specific referral to a service provider; in Portland, 70 percent of contacts resulted in no specific referral. As discussed below, although these "nonspecific" contacts did not include identifiable referrals, they did involve substantive and meaningful contact in which case managers provided support or delivered counseling.

Outcomes related to drugs and crime

Tables 5 and 6 summarize behavioral outcomes for the three treatments. Table 5 presents results for Washington, D.C., and Table 6 for Portland, Oregon. For each of the three treatments, the tables report average outcomes for behavioral prevalence at baseline, 3 months, and 6 months for cases that entered the analysis. Prevalence means during the 3-month period before each interview, except rearrested since baseline. They also show the total sample size for each outcome. Sample sizes vary due to variations in response rates or in the subsample to which the outcome measure applies. The baseline figures are for all participants included in either the 3- or 6-month samples. The other two prevalence measures apply to respondents who answered the 3-month or 6-month interview, respectively.

With one exception (use of condoms by sexually active participants in Portland), participants in all three treatments reported statistically significant, and often dramatic, changes in behavior after enrollment. We have no way to determine how much of this overall change can be attributed to the treatments. Although study participants were randomly assigned to one of the three treatments, they had first volunteered to take part in the study, and we have no control group that would allow us to estimate how they would have behaved in the absence of any treatment. Further, they were subject to other potentially important interventions. In particular, many were under criminal justice agency supervision during all or part of the followup period, and a substantial minority in all three treatments reported participation in various other drug treatment programs.

For example, although it seems unlikely that the videotape shown to participants in the "video" treatment accounted for a substantial part of the drop in the percent of participants reporting heavy drug use from 86 percent at baseline to 28 percent at 3 months after enrollment, we cannot say for sure that it contributed nothing. Even so, given the modest nature of the video intervention, it seems reasonable to regard the outcomes for this group as reasonably close to those of a no-treatment control group— remembering that, in this case, "no treatment" frequently includes the effects of supervision by criminal justice agencies, among other things.

We can, however, compare outcomes among the three treatments and thus determine the effects of the differences in the additional interventions offered by these. We tested differences among the treatments in two ways. First, we tested case management against the other two treatments combined. This reflects our special interest in the coordinated counseling and referrals offered by case management and the fact that the outcomes for those assigned to video or referral were often quite similar. Statistical testing was based on a multivariate probit model that assumed correlation across events. We used one-sided tests, testing only the hypothesis that outcomes at 3 and 6 months were not better for those assigned to case management than for those assigned to the other two interventions. The results of these comparisons are indicated in tables 5 and 6 by the symbols next to the case management entries. A symbol of # indicates that the average outcome for those assigned to case management was significantly better than the average outcome for those assigned to the other two treatments, using a 0.05 test level. A symbol of ## indicates significance at a 0.10 test level. All estimates are based on a model that adjusts for several covariates and for differences in baseline behavior.

In addition to this overall comparison, we compared case management with each of the other interventions individually. The results of these comparisons are indicated by the symbols next to the entries for the other two treatments. The symbol * by an entry for referral or video at 3 or 6 months is used to indicate that outcomes for case management were significantly better than those for that treatment (at that point in time) using a 0.05 test level; the symbol ** is used to indicate that they were significantly better at a 0.10 test level (again, using a one-sided test).

The proportion of participants reporting heavy drug use declined dramatically in all three treatments in both sites. There were also substantial increases in the proportion reporting participation in drug treatment programs. In addition, in Washington, a significantly smaller proportion of those assigned to case management reported heavy drug use during the 3 months before both the 3- and 6-month interviews, and a significantly greater proportion reported participation in drug treatment programs in the 3 months before the 3-month interview. In Portland, a significantly smaller proportion of those assigned to case management reported heavy drug use during the 3 months before the 3-month interview, but there was no significant difference among the three treatments in the incidence of heavy drug use before the 6-month interview, or any difference in reported participation in drug treatment programs during the 3 months before either the 3- or 6-month interview. However, these self-reports of drug treatment are inconsistent with public treatment records (not shown in the table), which showed that 19 percent of the case management participants, 12 percent of those who received referrals, and 14 percent of those who only saw the video had entered public drug treatment programs within 6 months of the baseline interview, suggesting that case management does increase entry into substance abuse treatment and that the self-reports are misleading. Public treatment records were unavailable for Washington, D.C.

Participants in this study—many of whom were under criminal justice supervision for some part of the 6-month followup period, although not by staff at the study sites—reported dramatic reductions in illegal activity from the prearrest period (see tables 5 and 6). For the 3-month period just before the 6-month interview, case management participants in both sites reported significantly less criminal behavior than other participants. Time spent in jail (in D.C. only) was significantly lower for case management participants than for other participants. In Portland, time spent in jail was lower for case management participants for the 3-month period before the 6-month interview, based on a comparison of case management outcomes with the two other interventions combined. In Washington, D.C., the reported reduction in criminal involvement was corroborated by CJS data showing that case management participants were significantly less likely to be rearrested than were other participants.[12] Comparable data were not available from Portland.

Outcomes related to HIV prevention

Without regard to intervention group, participants in both Portland and Washington reported reduced needle use, reduced needle sharing, and increased needle cleaning (see table 6). Only in Portland was case management associated with an additional decrease in self-reported needle use and needle sharing during the 3 months before the 3-month interview. There were no important differences in the rates at which those who persisted in sharing needles cleaned their drug paraphernalia. In Washington, there were no discernible differences across the three treatment groups, and so few participants shared needles that analysis of the needle-cleaning behavior was impractical.

In both cities, significantly fewer participants across all interventions groups reported multiple sexual partners between baseline and followup interviews (see tables 5 and 6); but discounting a single significant effect in Portland, the decrease was no greater for case management clients than for other participants. Among all participants, the trend toward increased condom use among sexually active participants reached statistical significance in Washington, D.C., but not in Portland. In Portland, all three groups used condoms at about the same rate at 3 and 6 months, although this represented a significant increased use of condoms at 3 months for the case management group.

Limitations of the Evaluation

Evaluation studies rarely are definitive, and this one is no exception. Self-reports of changed behavior play an important role in this evaluation. Unfortunately, self-reports may be inaccurate. Respondents may exaggerate socially desirable behaviors (such as entering drug treatment and using condoms) and understate socially undesirable ones (such as injection drug use). Some of the improvement reported by participants, no matter what their intervention assignment, may have reflected such biases. Improvements specifically associated with case management—reductions in drug use and criminal behavior and increases in substance abuse treatment—may have been different than they appeared to be.

Besides possible inaccuracy in self-reports, two additional features of this study associated with the experimental design limit its generalizability. To collect data, the evaluators paid participants to answer a series of three interviews. Although no client received payment in exchange for accepting case management services, there is no way to know for sure whether initial recruitment would have been equally successful without the stipend for the interview. Also, this study evaluated specific variations of case management in two specific settings at a single point in time. The extent to which these findings generalize to other populations of arrestees in other settings at other times is speculative, although there is no reason to suppose that similar results would not arise elsewhere.

Improving the Delivery of Case Management

Despite these limitations, case management seemed to effect improvements in behaviors that are often considered intractable in a population resistant to ameliorative interventions, particularly interventions in which participation is voluntary (Inciardi et al., 1994). These results suggest that additional refinements, building on what appear to be the strengths of this approach and addressing apparent weaknesses, might effect still greater improvements in outcomes related to drug abuse and criminality. These are discussed below.

The client-case manager relationship in the promotion of behavior change

Extensive analyses were performed to determine whether case management worked by increasing utilization of community services. Notwithstanding the imprecision of the measurement instrument and assessment schedule, referrals offered to individual case management clients did not correlate specifically with needs they reported on the assessment interviews, nor was utilization of services other than drug treatment measurably greater for case management clients than for participants assigned to the less intense comparison group interventions. It is possible that case management clients already were well aware of the existence and limitations of services in their community, which would have reduced the specific impact of referrals. Other projects working with comparable populations (e.g., Finn et al., 1992; Falck, Ashery, Carlson et al., 1994; Falck, Carlson, Price et al., 1994) to provide case management suggest that the availability (or absence) of relevant services is well known to members of the target population.

Even if it did not increase clients' access to services, case management may have increased retention in services that participants accessed or enhanced the effectiveness with which clients utilized services (Teitelbaum and Gross, in preparation). Data on retention in drug treatment and associated improvements among case management participants in this study are consistent with that hypothesis. At least for this population, the most important element of case management apparently was not the provision of referrals. Possibly case management supported continued and effective utilization of services. First, case management may have helped clients solve practical problems associated with receiving services such as transportation and child care. Second, case management may have taught participants the client role: habits such as keeping appointments, talking about personal matters with a service provider, and accepting advice (Brown and Needle, 1994).

Long-term, indepth case history interviews of case management clients contain ample testimony of the importance of the case manager's role in motivating self-improvement. Important dimensions of that role seem to include providing support during periods of crisis, enhancing clients' self-esteem by emphasizing their past accomplishments and current strengths, and encouraging sustained improvement without engaging in blaming or recrimination when clients relapse or recidivate. Even clients who did not participate heavily expressed favorable attitudes about the relationship with their case managers, as exemplified in statements from clients at the Washington, D.C., site (*I* indicates infrequent participation [less than five contacts] and *F* indicates frequent participation [more than 24 contacts]):

- "Hearing myself talk helped me see what I'd become."(I)

- "Luckily I qualified for the . . . caseworker. I would call if I got depressed. They gave me the push for my treatment." (F)

- "_____ persisted. She didn't stop in spite of me rebelling." (I)

- "_____ was a friend, a good listener, an inspiration." (F)

- "I knew someone was there to help me do better—someone who really cared." (I)

- "She didn't even know me, but she showed that she cared. She was like a mother to me—when I was feeling down, I could turn to her." (F)

Respectful treatment of clients (including seemingly trivial but telling gestures such as making eye contact and shaking hands) led clients to feel that case managers saw them as more worthy than just "dumb dope fiends."

Coerced versus voluntary participation in services

The parent agencies that implemented these programs at first doubted that volunteers would participate in case management without criminal sanctions for noncompliance or substantial material incentives. However, a majority of those assigned to case management did participate, and open-ended case study interviews suggest that the absence of criminal sanctions actually strengthened the rapport between clients and providers by helping case managers convince clients of their authentic concern. Case managers were perceived as unlike "typical" parole or probation officers, as "friends" who could be trusted with information about illicit behavior. An identity as new agencies, without overt associations with the CJS, apparently increased the appeal of these programs.

Agencies with reputations for conducting CJS supervision may have difficulty creating the separation effected in this study between the coercive components of the CJS and the supportive, trust-related aspects of case management without sacrificing established interagency linkages and credibility among community service agencies. The combination of capabilities and experiences—working in the context of the CJS, familiarity with addiction and drug treatment modalities and services, experience working with the affected demographic groups, and knowledge of HIV prevention strategies—may be difficult to find in an existing agency. Creating new organizations inescapably will impose startup costs.

Harm reduction in the criminal justice context

As a condition of funding in 1989, the U.S. Department of Justice prohibited needle hygiene instruction and required promotion of sexual abstinence or mutual monogamy in favor of instruction on condom use. Given these injunctions, the only way to provide sustained HIV-risk reduction counseling and education was by linking participants to community HIV prevention programs through case management. Consistent with the CJS emphasis on refraining from illicit behavior, and a program philosophy that other behaviors would not change until addiction was addressed, staff at both study sites emphasized elimination of drug use as a means of reducing HIV-risk behavior and succeeded in reducing heavy drug use. As a strategy for HIV prevention, case management may need to address injection-related and sexual risk behavior more directly than simply referring clients to outside counselors. An alternative approach—which was not evaluated here—would be to have case managers themselves counsel persistent needle users in how to clean injection equipment and to discuss condom use with all their clients.

Given adequate training, sufficient direction, and consistent supervision, case managers may be well positioned to provide sustained counseling about injection-related and sexual risk reduction. Although the integrity of the CJS remains a concern, this study established that a clear boundary can be maintained between case management and CJS supervision. Such a separation should allow a relaxation of restrictions so that future projects conducted in conjunction with the CJS can provide a more direct HIV/AIDS intervention.

Such HIV prevention counseling would need to be individualized. Because the case management process leads to a comprehensive picture of clients' life situations, crafting such tailored HIV counseling and education would be possible. Thus, a case manager working with a client attempting to remain abstinent from drugs would reinforce that intention; the case manager would not emphasize instruction in needle hygiene or access to needle exchange programs unless there appeared to be a significant likelihood of relapse. In contrast, for clients who were not motivated to reduce their injection drug use, case managers might continue to emphasize the possibility of drug treatment while ensuring that these clients are aware of methods to prevent transmission of HIV via injection equipment. With regard to sexual risk reduction, similarly staged and nuanced messages might be delivered, in which case managers affirm that abstinence or mutual monogamy with an uninfected partner are the surest means to prevent sexual exposure to HIV, but counsel clients with multiple unknown or high-risk partners about how to use condoms effectively.

Targeting the duration and recipients of case management

An arbitrary increment of 6 months may not be the optimal duration for this service. For some clients the period is too long; for others, too short. An unexpectedly high proportion of participants in long-term case study followup interviews reported a history of psychiatric hospitalization or other markers of mental illness. This was consistent with the large proportion of program participants with high scores on the screen for mental disorders used in the baseline interview (over one-third in Portland and almost one-half in Washington, D.C.).[13] The long-term followup interviews with these clients suggested that they may require long exposure to intensive case management, perhaps combined with professional mental health services, in order to achieve any meaningful improvement. Other clients, well served by a relatively brief initial period of intensive case management to help them get back on their feet, may benefit from further case management services on an as-needed basis thereafter.

In addition to a more flexible and responsible method for assessing the appropriate duration of service provision, future projects might consider developing methods to focus resources on those clients likely to benefit most from case management. Qualitative findings suggest that clients who benefit most from intensive case management are those who are motivated to change, who have been able to sustain prosocial and stable lifestyles at least for intermittent periods in the past, and who lack current access to social support. Development of an effective screening method to identify such clients would contribute to more efficient use of limited resources.

Emphasis on training and supervision

Several factors suggest that training and clinical supervision of service delivery staff need to be emphasized in nontraditional programs. Problems attributable to inadequate training and supervision include unevenness in the delivery of services, inadequate attention to termination, and episodes of dysfunctional staff behavior (e.g., bickering, hostility, and inappropriate personal relationships between staff and clients). Additional challenges for training and supervision will be imposed by expanding the role of case manager to encompass HIV prevention.

Case managers cannot be expected to provide effective HIV prevention counseling without substantial training on routes of HIV transmission, techniques of HIV prevention education and counseling, and strategies for counseling clients with HIV infection. With relatively inexperienced or unevenly trained staff, projects must provide substantial training on clinical issues in the client-case manager relationship, especially given the emphasis on counseling as an adjunct to the assessment, referrals, and advocacy functions of case management, and the proposed expansion of these programs to accommodate needle hygiene and sexual risk reduction education.

Many case management clients were surprised and disappointed when told that their case manager was closing their cases and ending their participation in the 6-month program. Both sites had official policies about client termination; however, these policies were carried out erratically. Termination must be integrated into treatment planning and thoroughly discussed with clients to ensure that gains made through program participation are not threatened by the abrupt end-of-service provision.

Testimony from case managers at both sites about staff dissension and burnout suggests that projects such as this must devote more attention and resources to supervision. Periodic review and revision of treatment or "action" plans and insistence on a carefully planned termination process are minimal requirements. Given the fluidity of the case manager role and the need for case managers to relax some of the traditional provider-client boundaries (e.g., by meeting with clients outside the office or by working to build rapport through some personal self-disclosure), clinical supervision assumes a very important role. Factors such as productivity and recordkeeping are valid indicators of job performance and appropriate considerations for managerial oversight of staff performance. Qualitative aspects of job performance (e.g., thoroughness of assessment, comprehensiveness of treatment planning, and appropriateness of referrals), which are subtler, should be addressed with equal vigor but in the context of a clinical supervision process that is clearly distinguished from administrative management.

Conclusion

Recommended improvements mentioned above in the design of case management services and program evaluation should not imply that the reported findings are disappointing. Rather, case management effected important positive outcomes for drug abuse, drug treatment, and criminal recidivism in the face of a variety of obstacles and limitations. Results from this study provide highly suggestive if not totally conclusive evidence that case management—as implemented in this experiment—can promote socially desirable behavioral change among members of a population who have proved recalcitrant to other behavioral interventions.

Case management as delivered in this project has similarities to existing publicly funded approaches to the target population but also differs in significant ways. Case management incorporated the drug counseling and referral elements of TASC programs but without the supervisory and coercive elements (including periodic urine testing). The absence of coercion appeared to enhance the relationship between case managers and clients. It incorporated elements of outpatient drug-free counseling, but addressed the reluctance of members of the target population to participate voluntarily in drug treatment by requiring case managers to make a special effort to engage their clients. In establishing rapport, case managers crossed some of the boundaries maintained in conventional counseling through limited amounts of self-disclosure and through direct expressions of emotional support. In contrast with TASC and outpatient drug treatment, this case management approach was more tolerant of clients who relapsed or recidivated.

Although the project was not altogether successful in changing targeted behaviors, the researchers' observations and qualitative findings suggest several ways in which this intervention might be improved to achieve targeted outcomes. For example, integrating HIV/AIDS interventions into the overall intervention would provide more effective coordination of case management services. Also, study findings indicate that more focused training and supervision could be helpful for case managers who do not have advanced training. The development of protocols that match clients and case managers in accordance with client needs and case manager skills may go a long way to improving the service delivery fit. Finally, the provision of a full-time resource person who is responsible for initiating and maintaining relationships with community resources may help free case managers to work more intensively with clients.

All this is speculation, of course. The study did not evaluate whether better supervision, improved training, and targeting of services could improve outcomes. But it is speculation that is based on observations of what did happen in these two programs and reasoning about how those outcomes might be improved. Although the study did not lead to a model program that should be held as a paragon, it provides a sound basis for believing that the form of case management practiced here can work and can be improved with suitable fine tuning.

Modest but suggestive outcomes from this project lay important groundwork for integrating a public health approach into a criminal justice setting. They help replace the rhetoric that "nothing works" with a more optimistic perspective that troubled lives can be changed for the better.

Notes

1. The 1991 National Household Survey on Drug Abuse (the Household Survey, conducted by the National Institute on Drug Abuse) questions a representative national sample about drug use, criminal activity, and arrests. An estimated 625,000 Americans admitted using cocaine on a weekly basis. About 76 percent of the weekly users admitted either to being arrested or to having committed a criminal offense (not including drug use) during the year prior to the survey. About 20 percent of all others admitted an arrest or a crime. Statistics are based on tabulations performed by Abt Associates. Also see Harrison and Gfroerer (1992).

Arrestees are more likely to use illegal drugs than people who are not arrested. The Drug Use Forecasting (DUF) system data (collected by the National Institute of Justice from 23 sites) indicate that during 1990 roughly 43 percent of arrestees tested positive for cocaine, 19 percent for marijuana, and 10 percent for opiates. Authorities consider the test used by DUF to be a conservative measure of recent drug use (see Mieczkowski et al., 1993; Visher and McFadden, 1991). A positive test means that the arrestee used cocaine within 2 or 3 days of the arrest. In contrast, Abt Associates' analysis of the 1991 Household Survey indicates that fewer than one-half of one percent of Americans admitted using cocaine on a weekly basis during the year prior to the survey. Even after accounting for underreporting in the Household Survey, arrestees are much more likely than other citizens to use drugs.

2. HIV prevention demonstration projects that had targeted out-of-treatment injection drug users suggested that paraprofessional staff could work effectively in nontraditional settings to counsel drug users directly and to link them to traditional drug treatment (Brown and Needle, 1994).

3. Case managers served in the role of referral specialist but were instructed that the one-time referral session required a more directive approach to referrals for HIV-related and drug-related services because there would be no opportunity to cultivate a sustained relationship with the client.

4. This intervention employed the same needs assessment and planning instruments used in the initial case management session but required a more prescriptive approach. Referral specialists placed more weight on their priorities for the client than on the client's perceived immediate needs, and they provided staged recommendations meant to structure help-seeking activities for a relatively long period (in contrast with the focus on short-term, client-centered goals established at early case management visits).

5. At one site, where this role was delegated to a paid consultant rather than to a staff member from Abt Associates, Abt staff sought the advice and consultation of agency officials responsible for managing the study to assist in identifying and selecting a suitable candidate. This did not guarantee an absence of friction in the monitoring relationship, but it ensured some degree of compatibility.

6. Candidates were recruited based on self-reports in Portland and through urinalysis testing in Washington, D.C. However, acknowledged drug use was an eligibility criterion at both sites.

Candidates were informed that a research project was seeking volunteers for a study of people with substance abuse problems and that candidates would be compensated for their participation in three interviews designed to understand their life circumstances and to determine how to help others like them. They also were told that they would be randomly assigned to one of the three research groups. (Arrestees in both sites were compensated for their participation in the baseline, 3-month, and 6-month interviews. In Washington, D.C., participants received $20 for each interview. In Portland, they received $15 for the baseline and 3-month interviews and $30 for the 6-month interview.)

Candidates were reassured that their participation was completely voluntary and would have no effect—helpful or harmful—on their pending court cases. To support that claim, a Federal Certificate of Confidentiality and court orders from each jurisdiction protected study-related information about clients from subpoena. After giving informed consent to participate, newly enrolled subjects answered questions which would help study staff to relocate them. Recruits provided information on the following: name, nickname, sex, race, age, date of birth, court identification number (later confirmed using court records), social security number, lawyer's name and phone number, home address and phone number, work address and phone number, alternate address and phone number, and other important contacts and their phone numbers. Those who agreed to participate were scheduled for a baseline interview, after which they received their intervention assignment.

7. CMRs measured intensity of referrals in three categories: The case manager (1) provided a name and phone number for a referral, (2) called a referral on behalf of a client to make an appointment for the client, and (3) accompanied the client to the referral site.

8. Specifically with regard to differential rates of followup across the three intervention groups, in Washington, D.C., case management participants composed 34 percent of those interviewed at baseline and 35 percent of those interviewed at the 3-month and 6-month followups. In Portland, Oregon, they composed 33 percent of those interviewed at baseline, 35 percent of those interviewed at the 3-month followup, and 34 percent of those interviewed at the 6-month followup.

9. Data not shown in table 1. DUF data and enrollment in other HIV-related and health programs in Portland, Oregon, also show underrepresentation of Latinos (Schlenger et al., 1992). The reasons are not clear but may relate to a lack of bilingual and especially bicultural staff and services, despite the presence of many monolingual Spanish-speaking migratory farm workers.

10. A significantly higher proportion of females received 24 or more contacts than did males (36 percent versus 23 percent, respectively; $p = .05$).

11. The difference in proportion of clients referred to drug treatment may have been related to differences in availability as well as in philosophy of the parent agencies. In Washington, D.C., effective utilization of any services and sustained behavior change was believed to depend upon addressing drug abuse which, by definition, affected all project participants. Portland staff may have made referrals to drug treatment only for those clients considered to have a severe drug abuse problem and/or to be prepared to engage in treatment.

12. Statistical significance ($p<0.10$) was based on a Kaplan-Meier survival model. For Portland, secondary data from the Oregon Justice Information Network showed no significant difference across intervention groups with regard to rearrest or reincarceration. However, those data appear to be incomplete: 16 percent of the study sample could not be matched with these records for the corresponding period despite the fact that everyone had at least one arrest (the arrest that made them eligible for this study) and should have had a record in the system.

13. The high prevalence on scales measuring mental disorders in this sample is consistent with other epidemiologic data on comorbidity of mental disorders and substance abuse. The Epidemiologic Catchment Area Survey of both community and institutionalized populations identified generalized anxiety disorder in 28 percent of substance abusers and major depressive and bipolar disorder in 26 percent (Regier et al., 1990). The criminal justice system may process a population in which rates of comorbidity are higher than in the general population.

References

Anthony, W.A., Cohen, M., Cohen, B., and Farkas, M. (1988). Clinical care update: The chronically mentally ill: Case management—more than a response to a dysfunctional system. *Community Mental Health Journal*, 24(3): 219–227.

Applebaum, R., and Austin, C. (1990). *Long-Term Care Case Management: Design and Evaluation.* New York: Springer.

Ashery, R.S. (1992). Case management community advocacy for substance abuse clients. *Progress and Issues in Case Management.* Rockville, Md.: National Institute on Drug Abuse Research Monograph 127.

Bandura, A. (1977). *Social Learning Theory.* Englewood Cliffs, N.J.: Prentice-Hall.

Bandura, A. (1984). Self-efficacy: Toward a unifying theory of behavioral change, *Psychological Review*, 84: 191–215.

Bloom, P., and Novelli, W. (1981). Problems and challenges in social marketing, *Journal of Marketing*, 45: 79–88.

Bonaguro, J., and Miaoluis, G. (1983). Marketing: A tool for health education planning, *Health*, (January/February): 6–11.

Brown, B.S., and Needle, R.H. (1994). Modifying the process of treatment to meet the threat of AIDS. *International Journal of the Addictions*, 29.

Centers for Disease Control and Prevention (1993a). *HIV/AIDS Surveillance Report*, February 1993.

Centers for Disease Control and Prevention (1993b). HIV prevention through case management for HIV-infected persons: Selected sites, United States, 1989–1992. *Morbidity and Mortality Weekly Report*, 42(23): 448–456.

Centers for Disease Control and Prevention (1994). Heterosexually acquired AIDS—United States, 1993. *Morbidity and Mortality Weekly Report*, 43: 155–160.

DeJong, W. (1991). On the use of mass communications to promote the public health. *Surgeon General's Workshop on Organ Donation: Background Papers*, Rockville, Md.: U.S. Department of Health and Human Services, Office of the Surgeon General.

Deren, S., Davis, R., Tortu, S., and Ahluwalia, I. (1993). Characteristics of female sexual partners. In Brown, B.S., and Beschner, G.M., eds., *Handbook on Risk of AIDS: Injection Drug Users and Sexual Partners*, Westport, Conn.: Greenwood Press, chapter 10.

Des Jarlais, D.C., and Friedman, S.R. (1987). HIV infection among intravenous drug users: Epidemiology and risk reduction. *AIDS*, 1: 67–76.

Falck, R., Ashery, R., Carlson, R., Wang, J., and Siegal, H. (1994). The use of social services by substance abusers, as cited in Ashery, R. (1994) Case management for substance abusers. *Journal of Case Management*, 3(4): 171–183.

Falck, R., Carlson, R., Price, S., and Turner, J. (1994). Case management to enhance HIV risk reduction among users of injection drugs and crack cocaine. In Ashery, R. (1994). Case management for substance abusers. *Journal of Case Management*, 3(4): 162–166.

Finn, P., Smith, C., Rhodes, W., Harrold, L., Cole, P., Teitelbaum, S., et al. (1992). Evaluation of an AIDS outreach and prevention program for injection drug users not in treatment and their sexual partners: The Patterson (New Jersey) Health Behavior Project. Cambridge, Mass.: Abt Associates Inc.

Gross, M., and Brown, V. (1993). Outreach to injection drug using women. In Brown, B.S., Beschner, G.M., and the National AIDS Research Consortium, eds., *Handbook on Risk of AIDS: Injection Drug Users and Their Sexual Partners*, Westport, Conn.: Greenwood Press.

Gross, M., DeJong, W., Lamb, D., Enos, T., Mason, T., and Weitzman, E. (1994). "Drugs and AIDS: Reaching for Help": A videotape on AIDS and drug abuse prevention for criminal justice populations. *Journal of Drug Education*, 24(1): 1–20.

Gross, M., Hunt, D.E., Cole, P., Harrold, L., Mason, M., Rhodes, W., and Smith, C. (1992). *AIDS Outreach to Pregnant Women and Their Children: Final Report to NIDA*. Cambridge, Mass.: Abt Associates Inc.

Hammett, T., Hunt, D., Gross, M., Rhodes, W., and Moini, S. (1991). Stemming the spread of HIV among IV drug users, their sexual partners and children: Issues and opportunities for criminal justice agencies. *Crime & Delinquency*, 37: 101–124.

Hammett, T., Hunt, D., Rhodes, W., Smith, C., and Sifre, S. (1992). *AIDS Outreach to Female Prostitutes and Sexual Partners of Injection Drug Users: Final Report to NIDA*. Cambridge, Mass.: Abt Associates Inc.

Harrison, L., and Gfroerer, J. (1992). The intersection of drug use and criminal behavior: Results of the National Household Survey on Drug Abuse. *Crime and Delinquency*, 38(1): 422–443.

Inciardi, J., Marton, S., and Scarpitti, F. (1994). Appropriateness of assertive case management for drug-involved prison releasees. In Ashery, R., ed., Case management for substance abusers. *Journal of Case Management*, 3(4):145–149.

Janz, N.K., and Becker, M.H. (1984). The Health Belief Model: A decade later. *Health Education Quarterly*, 11: 1–47.

Lefebvre, R., and Flora, J. (1988). Social marketing and public health intervention. *Health Education Quarterly*, 15: 299–315.

Leigh, B.C., and Stall, R. (1993). Substance use and risky sexual behavior for exposure to HIV. *American Psychologist*, 48(10): 1035–1045.

Leukefeld, C.G., Battjes, R.J., and Pickens, R.W. (1991). AIDS prevention: Criminal justice involvement of intravenous drug abusers entering methadone treatment. *Journal of Drug Issues*, 21: 673–683.

Martin, S.S., and Scarpitti, F.R. (1993). An intensive case management approach for paroled IV drug users. *Journal of Drug Issues*, 23(1): 43–59.

McGuire, W. (1984). Public communication as a strategy for inducing health-promoting behavioral change. *Preventive Medicine*, 13: 299–319.

McLellan, A.T., Arndt, I.O., Metzger, D.S., Woody, G.E., and O'Brien, C.P. (1993). The effects of psychosocial services in substance abuse treatment. *Journal of the American Medical Association*, 269: 1953–1959.

Mieczkowski, T., Landress, H.J., Newel, R., and Coletti, S.D. (1993). *Testing Hair for Illicit Drug Use*. Washington, D.C.: National Institute of Justice.

National Institute of Justice/Drug Use Forecasting Program (1993). *Drug Use Forecasting: Third Quarter, 1992*. Washington, D.C.: National Institute of Justice.

Regier, D.A., Farmer, M.E., Rae, D.S., Locke, B.Z., Keith, S.J., Judd, L.L., and Goodwin, F.K. (1990). Comorbidity of mental disorders with alcohol and other drug abuse. *Journal of the American Medical Association*, 264: 2511–2518.

Rehony, K., Frederiksen, L., and Solomon, L. (1984). Marketing principles and behavioral medicine: An overview. L. Frederiksen, L. Solomon, and K. Brehony, eds., *Marketing Health Behavior: Principles, Techniques, and Applications*. New York: Plenum.

Ridgely, M.S., and Willenbring, M.L. (1992). Application of case management to drug abuse treatment: Overview of models and research issues. *Progress and Issues in Case Management*. Rockville, Md.: National Institute on Drug Abuse Research Monograph 127.

Rothman, D. (1971). *Conscience and Convenience: The Asylum and Its Alternatives in Progressive America*. Boston: Little, Brown.

Rothman, J. (1991). A model of case management: Toward empirically based practice. *Social Work*, 36(6): 520–528.

Schlenger, W.E., Kroutil, L.A., Roland, E.J., and Dennis, M.L. (1992). "National Evaluation of Models for Linking Drug Abuse Treatment and Primary Care: Descriptive Report of Phase One Findings." Research Triangle Park, N.C.: Research Triangle Institute.

Siegal, H.A., and Cole, P.A. (1993). Enhancing criminal justice based treatment through the application of the intervention approach. *Journal of Drug Issues*, 22:131–142.

Siegal, H.A., ed. (1994). Special issue on the application of case management in the treatment of drug and alcohol abuse. *Journal of Case Management*, 3(4).

Steel, E., and Haverkos, H.W. (1992). Epidemiologic studies of HIV/AIDS and drug abuse. *American Journal of Drug and Alcohol Abuse*, 18(2): 167–175.

Stimson, G. (1992). Drug injecting and HIV infection: New directions for social science research. *The International Journal of the Addictions*, 27(2): 147–163.

Teitelbaum, M., and Gross, M. Linking HIV-related primary care and substance abuse services: The limitations of case management. (Under review at Health Resources and Services Administration, to be submitted to *AIDS and Public Policy Journal*.)

Turner, C.F., Miller, H.G., and Moses, L.E. (1989). *AIDS: Sexual Behavior and Intravenous Drug Use*. Washington, D.C.: National Academy Press.

Visher, C., and McFadden, K. (1991). *A Comparison of Urinalysis Technologies for Drug Testing in Criminal Justice*. Washington, D.C.: National Institute of Justice.

Weil, M., Karls, J.M., and Associates (1985). *Case Management in Human Service Practice*. San Francisco: Jossey-Bass.

Weissman, G., and the National AIDS Consortium (1991). AIDS prevention for women at risk: Experience from a national demonstration research program. *The Journal of Primary Prevention*, 12(1): 49–63.

Wexler, H.K., Magura, S., Beardsley, M.M., and Josepher, H. (1994). ARRIVE: An AIDS education/relapse prevention model for high-risk parolees. *International Journal of the Addictions*, 29: 361–386.

Wish, E.D., O'Neil, J., and Baldau, V. (1990). Lost Opportunity to Combat AIDS: Drug Abusers in the Criminal Justice System. In Leukefeld, C.J., Battjes, R.J., and Amsel, Z., eds., *AIDS and Intravenous Drug Use: Future Directions for Community-Based Prevention*. Rockville, Md.: National Institute on Drug Abuse Research Monograph 93.

Figure
and
Tables

Figure 1: Process for recruitment, randomization, intervention, and data collection

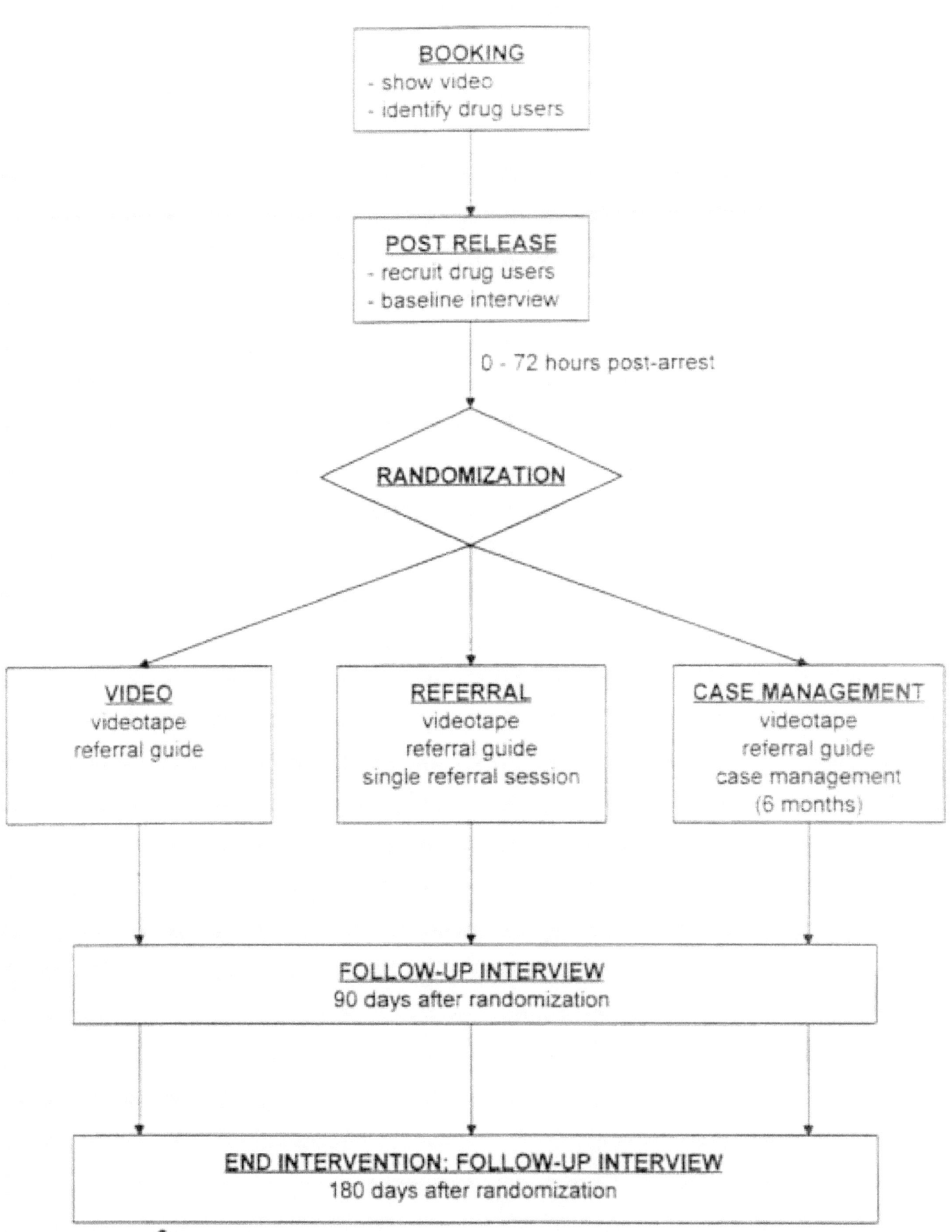

Table 1. **Participant Characteristics**

Characteristic	Washington, D.C.		Portland, Oregon	
	Baseline Participant Sample (n = 673)	**CJS Population**[1] (n = 57,960)	**Baseline Participant Sample** (n = 696)	**CJS Population**[2] (n = 21,445)
Male	74%	84%	74%	80%
Female	26%	16%	26%	20%
Black	95%	81%	34%	25%
White	3%	18%	51%	57%
Other/unknown	2%	.4%	15%	18%
18–19 Years	1%	12%	7%	14%
20–29	30%	39%	33%	35%
30–39	50%	31%	41%	32%
40–49	16%	12%	17%	14%
50 or Older	3%	6%	2%	5%
		(n = 7,585)[3]		(n = 1,141)[4]
11 Years or Less	44%	43%	37%	35%
High School/GED	52%	42%	59%	59%
Bachelor's Degree	4%	15%	4%	6%

Note: Some percentages do not sum to 100 percent due to rounding.

Sources: [1] D.C. Metropolitan Police Department (preliminary figures)
        [2] Oregon Uniform Crime Reporting
        [3] Pretrial Services Agency of D.C.
        [4] Drug Use Forecasting System (1990)

Table 2. **Drug Use and Injection During the Month Before Baseline**

Mutually Exclusive Drug Use/Drug Injection	Washington, D.C. (n = 673)	Portland, OR (n = 696)
Heavy Heroin (≥4 days/week) also using cocaine/crack daily or weekly injecting drugs	**19%** 66% 84%	**13%** 57% 95%
Heavy Cocaine/Crack (≥4 days/week) injecting drugs	**28%** 7%	**25%** 55%
Weekly Cocaine/Crack (≤4 days/week) injecting drugs	**35%** 6%	**30%** 44%
Heavy Alcohol also using cocaine/crack infrequently injecting drugs	**4%** 89% 4%	**8%** 69% 49%
Injecting Drugs injectors sharing injection equipment	**21%** 23%	**51%** 39%

Participants were classified according to six drug use categories:

· *Heavy heroin users* take heroin 4 or more days per week. Typically, one or two dime bags are used per session. Most users take heroin one to four times per day, although a few claim more frequent usage. At baseline, 75 percent had used some cocaine in the past month, apparently in combination with heroin. In terms of money spent, heroin is the dominant drug.

· *Heavy cocaine/crack users* consume cocaine or crack 4 or more days per week. They typically use one or two rocks or bags per session, usually costing $10 or $20 per unit. Sometimes a $50 rock is used. Heavy cocaine/crack users rarely consume other drugs at a high rate, except alcohol. In Portland, this category includes 3 percent who use amphetamines (but not necessarily cocaine/crack) on 4 or more days per week.

· *Weekly cocaine/crack users* take cocaine or crack each week but on fewer than 4 days per week. The $10 and $20 rock or bag is the typical dosage; some are binge users and spend more than $100 per session. The upper end of this category (3 days per week, especially with heavy use per session) is almost indistinguishable from the heavy cocaine/crack use category. In Portland, this category includes 2 percent who use amphetamines (but not necessarily cocaine/crack) weekly but on fewer than 4 days per week.

· *Heavy alcohol users* drink at least 4 days per week and consume at least three drinks per session. At baseline, 81 percent of heavy alcohol users in Washington, D.C., and 67 percent in Portland also used another drug heavily.

· *Infrequent cocaine/crack users* take cocaine or crack (3 percent in Portland used amphetamines but not necessarily cocaine or crack) less frequently than 4 days per month. Some spend more than $100 per session.

· A small residual category does not seem to use any of the above drugs, or any other drug, very heavily.

Because heavy (frequent) drug users were the target population, the last two categories were collapsed. The usage categories were ranked as above. When usage categories overlapped, the participant was classified into the highest category.

Table 3. **Self-Reported Need for Social Services During 3 Months Before Baseline**

Need	Washington, D.C. (n = 673)	Portland, Oregon (n = 696)
Health	47%	54%
Psychological Counseling	44%	37%
Psychiatric Care*	36%	47%
Housing	29%	49%
Social Support	24%	35%
Without Full-Time Employment	82%	84%
Monthly Income: $300 or less $301 to $799 $800 or more	35% 31% 34%	38% 32% 30%

*Evidence of mental illness, based on Referral Decision Scale.

Table 4. **Distribution of Case Management Contacts**

Number of Contacts during 6-month intervention period	**Percent of Participants Assigned to Case Management**	
	Washington, D.C. (n = 228)[*]	**Portland, OR** (n = 217)*
1–6 contacts	20%	42%
7–12 contacts	18%	23%
13–18 contacts	18%	12%
19–23 contacts	18%	6%
24 contacts or more	26%	17%

* Of 229 participants assigned to case management, one person had no contact with a case manager at the Washington, D.C. site; 13 of 230 participants in Portland, Oregon, also had no case manager contact.

Table 5. **Self-Reported Behavior and Case Management Impact, Washington, D.C.**

Target Behavior		Prevalence During 3 Months Before Interview		
		Baseline	3 months	6 months
Substance Abuse Related				
Heavy Drug Use				
	Case Management	85%	23% ##	17% #
	Referral	87%	29%	24% **
	Video	86%	28%	27% *
	Number of Cases	(612)	(574)	(568)
In Treatment				
	Case Management	13%	38% #	29%
	Referral	12%	28% *	24%
	Video	14%	31%	25%
	Number of Cases	(610)	(571)	(562)
Needle Use				
Used Needles				
	Case Management	20%	10%	5%
	Referral	19%	8%	6%
	Video	23%	9%	7%
	Number of Cases	(612)	(574)	(568)
Shared Needles				
	Case Management	3%	1%	1%
	Referral	5%	1%	1%
	Video	6%	2%	2%
	Number of Cases	(612)	(574)	(568)
Cleaned Needles+,§				
	Case Management			
	Referral			
	Video			
	Number of Cases		(5)	(5)
Sexual Risks				
Multiple Partners				
	Case Management	29%	16%	11%
	Referral	27%	16%	13%
	Video	28%	16%	12%
	Number of Cases	(612)	(571)	(568)
Never or Only Sometimes Used Condoms++				
	Case Management	49%	47%	47%
	Referral	51%	50%	57% **
	Video	51%	41%	42%
	Number of Cases	(381)	(329)	(289)
Crime Related				
Admitted to Criminal Behavior				
	Case Management	62%	19%	10% #
	Referral	61%	20%	17% *
	Video	61%	15%	15% **
	Number of Cases	(612)	(574)	(568)
Spent Time in Jail+++				
	Case Management		15% #	18% ##
	Referral		20% **	22% **
	Video		20% *	23% **
	Number of Cases		(563)	(548)
Rearrested Since Baseline++++				
	Case Management		12%	18% ##
	Referral		19%	27%
	Video		15%	21%
	Number of Cases		(671)	(671)

Case management is statistically different from the other two interventions combined at $p<0.10$ (##) or at $p<0.05$ (#) in a one-tailed test of significance. Case management is statistically different from the specified intervention at $p<0.10$ (**) or at $p<0.05$ (*).

+ Applies only to those who shared needles.

++ Applies only to those who had sexual partners.

+++ No baseline was recorded for this variable.

++++ Based on an analysis of criminal records rather than self-reports; the statistical test was limited to case management compared with the other two interventions.

§ This analysis is not available for Washington, D.C., due to small N.

Table 6. **Self-Reported Behavior and Case Management Impact, Portland, Oregon**

Target Behavior		Prevalence During 3 Months Before Interview		
		Baseline	3 months	6 months
Substance Abuse Related				
Heavy Drug Use				
	Case Management	80%	29% #	28%
	Referral	76%	36% **	26%
	Video	72%	36% **	27%
	Number of Cases	(590)	(513)	(561)
In Treatment§				
	Case Management	24%	38%	38%
	Referral	26%	40%	36%
	Video	19%	40%	42%
	Number of Cases	(587)	(510)	(558)
Needle Use				
Used Needles				
	Case Management	54%	24% #	21%
	Referral	52%	33% *	24%
	Video	47%	32% *	25%
	Number of Cases	(590)	(513)	(561)
Shared Needles				
	Case Management	19%	7% #	5%
	Referral	21%	13% *	8%
	Video	17%	11% *	6%
	Number of Cases	(590)	(513)	(561)
Cleaned Needles+				
	Case Management	70%	89%	75%
	Referral	22%	43%	50%
	Video	67%	77% *	80%
	Number of Cases	(43)	(36)	(19)
Sexual Risks				
Multiple Partners				
	Case Management	19%	10%	10%
	Referral	22%	11%	11%
	Video	18%	15% **	13%
	Number of Cases	(589)	(513)	(561)
Never or Only Sometimes Used Condoms++				
	Case Management	69%	59% **	58%
	Referral	57%	57%	56%
	Video	61%	60% **	63%
	Number of Cases	(325)	(252)	(257)
Crime Related				
Admitted to Criminal Behavior				
	Case Management	69%	31%	22% ##
	Referral	63%	33%	26%
	Video	67%	29%	27% **
	Number of Cases	(590)	(513)	(561)
Spent Time in Jail+++				
	Case Management		26%	23% ##
	Referral		27%	27%
	Video		24%	27%
	Number of Cases		(485)	(544)
Rearrested Since Baseline++++				
	Case Management			
	Referral			
	Video			
	Number of Cases			

Case management is statistically different from the other two interventions combined at p<0.10 (##) or at p<0.05 (#) in a one-tailed test of significance. Case management is statistically different from the specified intervention at p<0.10 (**) or at p<0.05 (*).

+ Applies only to those who shared needles.

++ Applies only to those who had sexual partners; behavior did not change from baseline.

+++ No baseline was recorded for this variable.

++++ This analysis is not available for Portland.

§ An analysis of treatment records showed that case management clients were significantly more likely to enter a publicly funded treatment program during the followup period.

Appendixes

Appendix A

How the Project Sites Implemented Case Management

Key features of case management as delivered at project sites are delineated below. Variations in project management and other site-specific factors meant that these components were not always fully implemented or consistently provided. Rather, taken together they represent a somewhat idealized version of the program that combines the features most fully realized at either or both project sites. The body of the report suggests possible refinements to the model described below that might enhance its success.

Staffing case management programs

Each of the two sites delivered case management services to approximately 225 clients, recruited over a 12-month period and seen for a period of 6 months each. The two sites each employed on average five case managers at any given time. Each also employed a project coordinator, a part-time (25 to 50 percent) social services coordinator, a part-time (25 percent) psychological/training supervisor, and a receptionist/administrative assistant.

Some, though not all, case managers had college degrees, a few had advanced degrees, and most had only limited professional experience delivering case management. Perhaps the most important qualification was high motivation to work with the target population and willingness to accept a full-time, but not permanent, position, funded on "soft money."

Developing a network of community resources

The social services coordinator contacted community agencies likely to be service providers for the target population. These were drug treatment programs, HIV counseling and testing projects and AIDS service organizations, employment and job training agencies, housing programs and shelters, health departments, community health and mental health centers, and offices providing public assistance (welfare, Aid to Families with Dependent Children, food stamps).

These initial contacts involved discussions with the responsible officials, requests for written letters of affiliation, and, in some cases, formal referral mechanisms with designated staff at the target agency. The social service coordinator asked the agencies to provide written materials about their services for a resource library at the project site.

The sites used these materials to develop a referral guide given to each study participant. At one site, the guide included photographs of the actual facilities so that clients would recognize the locations and might be less intimidated about visiting them.

The social services coordinator followed up with referral agencies to determine whether clients referred to them had made contact (after providing appropriate releases of information) and informed case managers. Conversely, case managers told the service coordinator and other case managers about service delivery agencies and programs. They reported on successes and problems making referrals to particular agencies, described barriers or difficulties encountered by clients, and called the attention of other staff to new services or programs or termination of previously available services.

Providing supervision and training

The psychological/training supervisor held weekly group case conferences that focused on cases that presented special problems or illustrated important approaches to case management. The psychological/training supervisor also met individually at least every other week with case managers to review their caseloads.

This individual also worked with the project coordinator to provide or arrange for staff training. Before they began to see clients, staff received training in the following key areas:

— HIV/AIDS epidemiology; counseling and testing; symptoms, medical referrals, and needs; and prevention principles.

— Manifestations and theories of drug abuse and drug treatment modalities.

— Techniques for delivering case management, including needs assessment, treatment planning, and working with provider agencies.

— Basic principles of mental health assessment.

— Multicultural sensitivity.

— Staff development issues (e.g., protecting confidentiality, establishing boundaries, stress management, personal safety, and ethical dilemmas).

The likelihood of staff turnover requires arrangements for continued inservice training and formal orientation of new staff who join the project after its inception. In these programs, case managers believed that continued training deserved more emphasis.

Establishing a relationship with the clients

The case managers' approach to establishing relationships with these clients—whose previous experiences of counseling or drug treatment tended to be limited, unsatisfying, or both—might be thought of as "befriending" clients. They struck a careful balance, being nonjudgmental or cautiously judgmental regarding illicit acts and enthusiastically supportive of prosocial and healthy behavior. They dealt with clients in a personal rather than official style (e.g., sending personal notes for birthdays).

Case managers sought to discover topics that interested the clients and could be a basis for engagement and ongoing interaction (e.g., sports and other recreational activities, music, children or family, and church). When clients were hard to reach or unresponsive, case managers were doggedly persistent but not punitive in their attempts to keep in contact. In addition to multiple letters and repeated telephone calls, they interacted with friends or family members as a way of establishing a network of linkages with clients who were difficult to reach directly.

In order to increase trust and avoid any implication that these efforts were a form of surveillance, case managers made sure that clients understood the program's strict separation from the criminal justice system. They explained that absolutely no information (other than overt plans to commit homicide or suicide) would be revealed to the authorities—not even information that might be beneficial to the clients—and obeyed that guideline rigorously. Clients were told to rely on other sources to document life changes that might benefit the adjudication of pending cases.

Also essential to the development of rapport between clients and case managers was flexibility. Case managers were encouraged to be flexible about meeting with clients outside the office but exercised this option only rarely; the case management process tended to be heavily office based. On the other hand, the predictable presence of case managers at the office provided another form of flexibility by ensuring that staff would be present if a case management client dropped in without an appointment. This flexibility proved to be one of the attractive features of the program for clients who participated actively.

Delivering case management

A baseline interview determined information about clients' core needs (e.g., drug treatment, housing, public assistance, and shelter). The interviewer gave a summary to the participant, who gave it to each case manager at the start of the first case management session. At that initial session, the case manager used a structured interview to learn about the client's situation, attitudes, and motivation regarding drug use and treatment; HIV prevention, counseling, and testing; and priorities regarding needs for assistance.

Using this information and baseline interview summary data, the case manager negotiated an initial *action plan* with the client. The case manager and client scheduled a followup appointment within 2 weeks—sooner if the client required crisis intervention or more immediate repeat contact.

Although case managers were told that they should keep in mind clients' needs for HIV prevention services, clients often had relatively little motivation to deal either with HIV or with their addiction. On the other hand, they often had a series of other needs they considered to be urgent. Typically these related to housing, employment, family, legal, or health crises. Initial case management referrals usually deferred attention to the HIV-related referrals in order to address these immediate needs, with the aim of helping the client to stabilize his or her life situation.

When clients did not appear regularly for appointments, case managers contacted clients on the telephone or by mail to keep them informed about developments in the community that might be beneficial to them and mailed special announcements (about jobs or training possibilities). The programs offered practical assistance together with referrals. For example, case managers helped clients complete paperwork for applications to receive services, wrote letters of introduction for food or shelter agencies that required a formal referral, and offered clients assistance in writing resumes.

Case managers often found it difficult to adhere to a formal treatment plan. Clients' chaotic lives, vulnerability, and multiple needs—together with limitations on available resources such as jobs or housing—meant that unanticipated crises might surface at any point or that unexpected factors would interfere with adherence to the routines of a service program to which they had been referred.

During the 6 months of service delivery, case management sessions continued to deal with the more conventional referral component of case management: an ongoing assessment of needs and the extent to which services had been successfully accessed. At the same time, the sessions with case managers also continued broader discussions of clients' life situations with a special emphasis on how clients were addressing drug use and any pending legal case.

One characteristic feature of counseling in the context of case management might be described as "reflective listening": case managers gave clients latitude to talk about the things they cared about, listened patiently, and asked questions to draw them out still more. Through these interactions clients identified needs they had not identified previously and, guided by supportive observations from case managers, they came to appreciate more fully their own strengths and inner resources.

Appendix B

"Drugs and AIDS—Reaching for Help": A Health Promotion Videotape for Criminal Justice Populations

This 27-minute videotape[1] uses a drama-based approach to convince inmates in jails, lockups, and booking facilities that they should adopt behaviors, including enrollment in drug treatment, to prevent the spread of acquired immunodeficiency syndrome (AIDS).[2] The videotape depicts both criminal sanctions for illicit drug use and the dire results of advanced human immunodeficiency virus (HIV) disease. Its real emphasis, however, is to encourage viewers to come to terms with both addiction and HIV prevention by drawing upon helping resources in their own communities. The congruence of general principles of health promotion with the testimony of recovering and active drug users who had "been there" suggested that such a positive theme would be most effective.

A brief synopsis of the videotape

The videotape presents three stories about people like those in the target audience—people whose lives have been seriously affected by drug abuse, who have experienced multiple brushes with the law, and who have in some way been affected personally by AIDS or infection with its causal agent, HIV. The stories tell how and why the protagonists turned their lives around, emphasizing the complementary roles of formal treatment and informal support from self-help programs, friends, and family in dealing with addiction and HIV prevention. Consistent with the emphasis on seeking help from others, the videotape concludes with a list of resources for finding assistance in one's own community.

The first story is told by "Shirley," an African-American woman in her early thirties who is HIV-infected but essentially healthy. Shirley has been clean for 2 years after almost two decades of drug abuse, recurrent arrests, and futile efforts to remain drug-free. The importance of love and support from Shirley's friend Liz and other members of her self-help group emerges in scenes in which Shirley discloses her fears of getting sick and "slipping" back into drug abuse, fears reinforced by the recent death of one of her closest friends, a white prostitute, from a drug overdose.

In the second story, "Carlos and Tina," a Latino couple with a young son, Dito, describe their struggle to come to terms with Carlos' AIDS diagnosis. They recount their fears that Tina will become infected, their heated but eventually successful negotiations over condom use, and their neighbors' growing understanding that they will not get the virus through casual contact. The story begins with a stark and shocking contrast between Carlos' robust good looks a few years earlier and the severe wasting caused by AIDS. At one point he remarks pathetically, "Even my tattoos have shrunk." Carlos' impending death is tragic, but he also expresses a father's hope that his son may have a better life, free of addiction, disease, and violence.

The third story depicts the relationship between a young African-American man, "B.J. Johnson," and his older counselor, Gregory Davis, also African American. Greg's role as B.J.'s mentor and role model began when B.J., at the completion of a 30-day, court-mandated drug treatment program, was assigned to Greg for counseling. B.J. belongs to a basketball league that Greg organized, which provides him and other young African-American men with a physical outlet and an appealing alternative to drug use. At one point we see B.J. and Greg in an intense group discussion about relapse and the associated risk of HIV exposure. Later, surrounded by drug-free friends at a dance, B.J. remarks of his new lifestyle, "Hey, it ain't so bad at all!"

The videotape uses the technique of photo-animation, with sequences of still photographs paced to convey action such as the slamming shut of a jail cell door, or to heighten emotion by fixing the viewer's attention on an arresting visual image such as the juxtaposition of B.J.'s hands shooting a basketball then shackled in handcuffs. Volunteers of all races—people who had been involved with drug abuse, spent time in jail, or become infected with HIV—agreed to be photographed for the videotape. Voices on the sound track are professional actors whose words are based on a composite of recorded interviews with former drug addicts, repeat offenders, sex workers, and others affected by crime, drugs, and HIV.

Health promotion principles

A criminal justice perspective might suggest dramatizing only the frightening consequences associated with illicit drug use and sexual behavior: incarceration, illness, and death. However, the National Research Council concluded that for HIV prevention the effectiveness of fear-inducing messages alone is "doubtful" (Turner et al., 1989). A threatening message may not motivate individuals to take action if the threat seems unlikely or remote in time. If overdone, such methods may paralyze individuals with fear, induce fatalism, cause them to derogate the credibility of the message, or lead them to deny their susceptibility (DeJong, 1991).

Several theoretical formulations that pertain to health behavior—the health belief model (Janz and Becker, 1984), social learning theory (Bandura, 1977, 1984), and the tenets of social marketing (Bloom and Novelli, 1981; Rehony et al., 1984; Lefebvre and Flora, 1988)—suggest that, to succeed, health education materials must be able to answer several questions for the target audience: What am I to do? Can I succeed in doing it? What will my friends think? What will I have to give up? What's in it for me?

The health belief model proposes that health-promoting behavior is fostered by a sense of **susceptibility** to the disease and of its **severity**, by a belief in the **efficacy** of the recommended action, and by **reduced barriers or obstacles** to executing the behavior. Also important is the **normative expectation** that the behavior is appropriate to one's culture and that one's peers have adopted it.

An important element of social learning theory is its emphasis on **self-efficacy**, the conviction that one is able to perform a certain behavior that will result in desired outcomes. Social marketing places particular emphasis on the **incentives or benefits** associated with the desired behavior. Applied to health behavior, this approach would associate risk reduction less with improved health than with universal human desires for acceptance, love, security, status, wealth, or beauty (Bonaguro and Miaoluis, 1983; McGuire, 1984).

Script development was based on the content of interviews with dozens of informants: current users, people in recovery, jail and prison inmates, and drug counselors. Fourteen focus groups in three different regions of the U.S. viewed and commented on a rough cut version of the videotape. This feedback was pivotal for editing and script revisions in the final version (Gross et al., 1994).

Application of health promotion principles in the videotape

Personal stories portray abandoning a criminal lifestyle as a positive step, one that restores physical and spiritual well-being, builds self-esteem, reinforces positive connections with family members, and sets the stage for further life-enhancing opportunities. It is often effective in a drama-based videotape to focus on a decision-point faced by one of the main characters. Shirley talks about how losing a baby while in jail because of cocaine, together with the discovery that she was HIV-infected, precipitated her quest for sustained treatment. B.J. describes his counselor Greg's influence on his decision to take treatment seriously.

Although the videotape is frank about the difficulties to be encountered in trying to free oneself of addiction, it emphasizes throughout that social support can be found and that success can be achieved. Especially important are the videotape's vivid and emotional renderings of the treatment process, particularly group meetings and self-help sessions, which may help viewers develop the conviction that they too can make treatment work. Presenting characters similar to the viewers themselves supports the belief that success in treatment does not require "superhuman" effort but is within reach with sufficient motivation and social support.

Many addicts deny their addiction or believe their drug use is under control. Shirley counters this belief when she recounts the unsuccessful struggle of her friend Janice to locate effective treatment, reflects on Janice's death from an overdose, and admits, "Sometimes I think I could slip too." She says that she thought drugs were fun at first but they "became a nightmare." Carlos reframes the "fun" of a drug-involved lifestyle as dangerous, frightening, and stressful. B.J., disillusioned, observes his former friends "doing crime, robbing their own mothers, doing anything for a hit." The videotape integrates the message that incarceration is a constant threat in the lives of many addicts. Gregory bluntly tells B.J. and his friends, "If you start using, you're going to get locked up. That's how it is."

Susceptibility to HIV through sexual exposure is addressed in each story. Shirley, who looks healthy and attractive, wonders if men would still be "hitting" on her if they knew she was HIV-positive. After Tina's ultimately successful struggle to persuade Carlos to use condoms, she considers her freedom from HIV infection a "miracle." B.J. confides that all the men in his group feel threatened by the risk of sexual transmission of HIV.

In contrast to the more immediate threats of incarceration or death by violence or overdose, drug addicts often think of dying from AIDS as a distant and abstract problem. Fear of death seems to be a less powerful threat than immediate disabilities from HIV-related illness—severe fatigue, chronic pain, emaciation, and dependency. The videotape vividly depicts the severity of such disability; audiences often gasp when they first see Carlos's emaciated body. But the videotape avoids conveying despair and futility. Carlos, for example, is determined to live as long as he can to help raise his son, and the closeness between father and son is apparent. Shirley contrasts a grim picture of "running"—an endless cycle of drug use, crime, punishment, and degradation—with her newfound pleasure in everyday things— playing with her friend Liz's baby, going to the circus, a bubble bath, and time with her friends.

The videotape acknowledges obstacles: treatment failures occur; effective treatment may be difficult to find; staying with a treatment program requires hard work. Conditions that maximize the chances of success, such as strong social support, are accentuated. B.J. describes his irregular attendance at groups initially and his growing involvement in response to Greg's encouragement. Shirley admits it would be easier to "stay home and watch TV" but understands that "I have to go [to meetings] and so I go."

Participation in groups and meetings and recreational activities with drug-free friends are depicted as a part of the fabric of a rewarding, drug-free life. Shirley and B.J. associate with attractive friends who support abstinence. B.J.'s excitement about playing on a drug-free basketball team is infectious. B.J., his girl friend, and other friends enjoy an evening at home and celebrate their sobriety at a drug-free dance. B.J. forcefully asserts, "If this is what they call being a sucker, a square, or lame, I don't mind."

A positive message

Drug treatment specialists and people in recovery have endorsed this approach, noting that it is essential to convey positive reasons for embracing a drug-free lifestyle and to give people the conviction that when they reach for help, recovery will be within their grasp. Many jail and prison officials have reacted favorably too, but not all. Some officials who viewed the pretest videotape discounted its approach and felt that what was needed instead were graphic portrayals of people suffering withdrawal symptoms in jail and hard-headed, no-nonsense messages that AIDS is a "death sentence" for illegal drug use. Criminal justice officials who are suspicious of this "nonpunitive" approach might reconsider: For promoting health behaviors it seems better to emphasize what people will gain, not just what they might lose.

Notes

1. The videotape can be ordered from the National Criminal Justice Reference Service, Box 6000, Rockville, MD 20849–6000, phone (800) 851–3420 or e-mail askncjrs@ncjrs.org for $14.00. Make your check payable to NCJRS and ask for NCJ 132940.

2. Other videotapes on AIDS prevention had already used a drama-based approach to promote risk reduction, such as "Olga's Story" and "Alicia" (both distributed by Modern Learning Aids, St. Petersburg, Florida), but productions using this approach had not been designed for persons under the supervision of the criminal justice system.

A detailed account of the implementation of this project, the evaluation methodology, and the results can be found in the final project report, *AIDS/HIV Education in Lockups and Booking Facilities,* by Michael Gross, William Rhodes, Catherine Conly, Tammy Enos, Stacia Langenbahn, Theresa Mason, and Linda Truitt. This report will be made available in late 1997 from the Fee-for-Service program of the National Criminal Justice Reference Service (see below).

For more information on the National Institute of Justice, please contact:

National Criminal Justice Reference Service
Box 6000
Rockville, MD 20849–6000
800–851–3420
e-mail askncjrs@ncjrs.org

To view or obtain an electronic version of *Case Management Reduces Drug Use and Criminality Among Drug-Involved Arrestees: An Experimental Study of an HIV Prevention Intervention* from the NCJRS Bulletin Board System, access the system in one of the following ways:
Direct dial through your computer modem: (301) 738–8895
(Modems should be set at 9600 baud and 8–N–1.)
Telnet to bbs.ncjrs.org
Gopher to ncjrs.org:71
For World Wide Web access, connect to the NCJRS Justice Information Center at:
http://www.ncjrs.org
If you have any questions, call or e-mail NCJRS.